MW00891932

LINCOLN SCHOOL
SCHOOL LIBRARY

Face-to-Face

with

The Chicken

Christian Havard

Photos by the Colibri and Jacana Agencies

Donated by
Reader To Reader, Inc.
Amherst, MA 01002

Charlesbridge

Chickens feast on earthworms and insects in the dewy grass.

At feeding time chickens peck with their beaks to pick up grain.

Using their talons, chickens scratch in the muck, looking for something to eat.

On the farm

Early in the morning, a rooster's crowing awakens farm animals. It is springtime, and the barnyard is full of activity. Pecking here and there, chickens move quickly to get as much food as they can. They scratch the ground with their feet to find more grain. Farmers fill troughs full of fresh water for chickens to drink.

Ground birds

Chickens preen, or clean, their feathers using their beaks. This preening rids chickens of parasites. Parasites are animals that get their food by living on other animals. Preening also helps chickens repair breaks in their feathers.

Chickens can fly, but they cannot go far. Usually they are too heavy, or their wings are too short and small for flying.

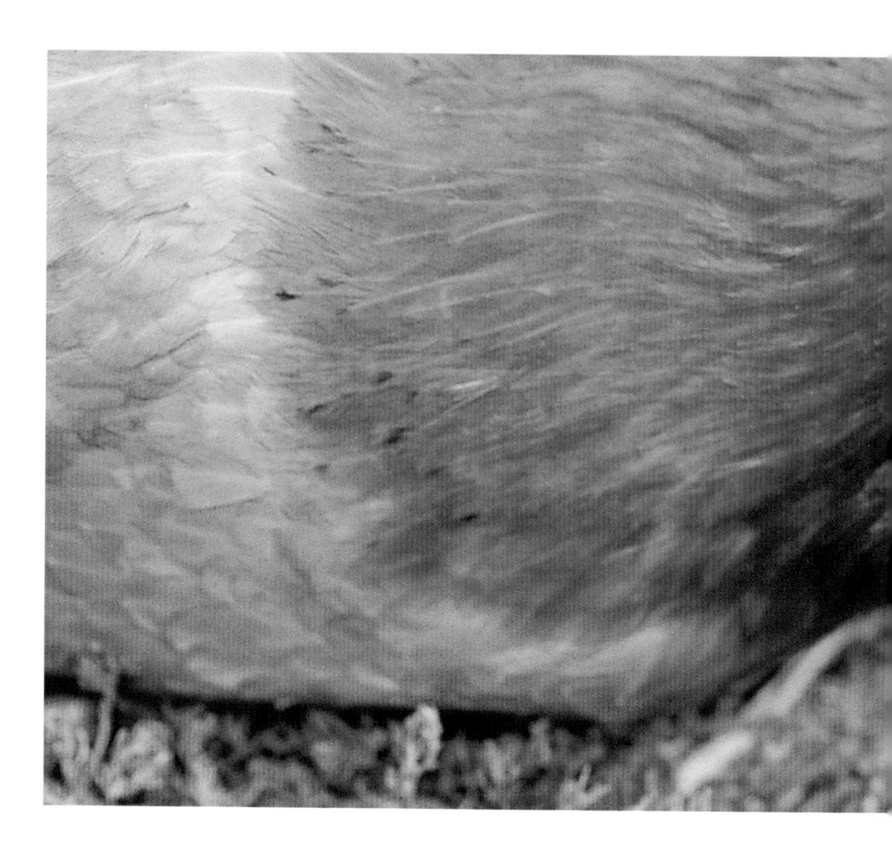

Each bird makes its own sound: roosters crow, hens cluck, and chicks peep.

A chicken's beak is hard and sharp, which allows chickens to pick up food easily. Chickens also use their beaks to defend themselves.

A chicken's feet, with their strong talons, are covered in scales.

To get rid of parasites, chickens rub their feathcrs in dry, dusty dirt.

Some chickens peck flies off other animals.

It's easy to identify the boss of a brood. The boss holds his or her head high and makes a lot of noise.

During the day, chickens that are lower in the pecking order stay apart from the rest to protect themselves from the pecking of other chickens.

Pecking order

In a brood, or group, of chickens, one is the boss. The boss is first in the pecking order and pecks all the other chickens without getting attacked. The chicken that is next in line pecks the others, but does not peck the boss. The pecking order allows the strongest chickens to get the best food and sleeping place. Each chicken pecks to find its place in the pecking order.

Dominant chickens will fight to get the food first with each sprinkling of grain.

Feeding time

When farmers toss grain to them, chickens scramble to be the first to eat. Chickens eat everything: bread, snails, or earthworms. They even swallow small pebbles. These pebbles move from a chicken's stomach to its gizzard, a pouch behind the stomach. Since chickens do not have teeth, the pebbles in their gizzard help crush food so it can be digested.

Mr. Rooster

Roosters, the male chickens, have a comb on the top of their heads and a wattle hanging from their throats. Hens are females, and they have a comb and wattle too, but theirs are smaller. When a rooster is ready to mate, he puffs up his feathers and dances around a hen. They mate, and the hen lays fertilized eggs. These eggs will produce chicks and will not be eaten.

Roosters have a long, sharp spur on each leg. They use these when fighting.

Roosters often have more colorful feathers than hens.

In order to stay boss, a rooster has to dominate young roosters, called cockerels, as well as the hens in the coop.

In a chicken coop there is usually 1 rooster for every 10 hens.

Hens sit on their eggs for 21 days until they hatch. They take good care to make sure their eggs are always kept warm.

Eggs that will be hatched should not be touched or gathered.

Eggs are made up of yolks and whites, protected by hard shells.

Hatching

Hens build their nests from pieces of straw, feathers, and down. For the next 12 days, they will lay 6 to 10 eggs at a rate of one per day. Hens lay eggs whether or not the eggs have been fertilized, but only fertilized eggs will produce chicks. Fertilized and unfertilized eggs look the same. Once they've laid a clutch, or group, of eggs, hens sit on them, only leaving to eat.

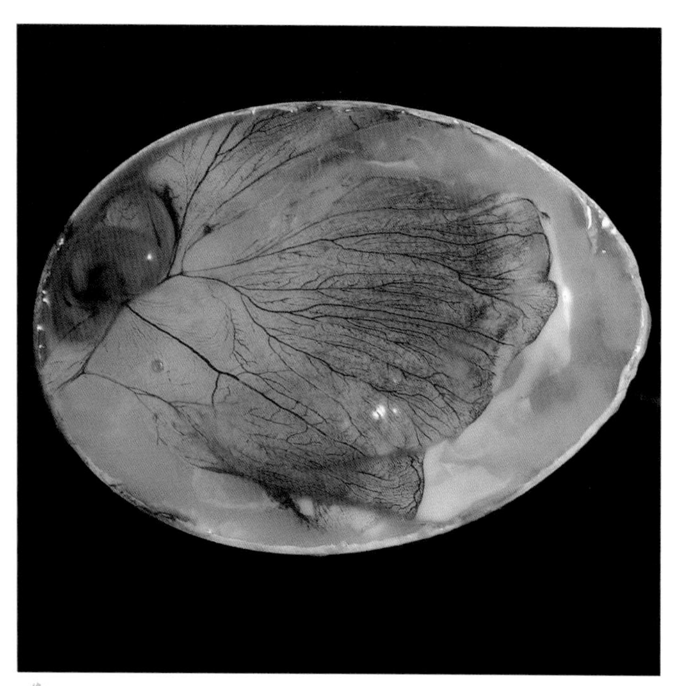

As embryos develop they are nourished by the white of the egg.

Chicks break their shells with the points of their beaks.

When chicks first hatch they are curled up and wet.

Once out of their shells, chicks dry off.

Chicks learn by watching their mother.

Hello, little chicks!

After a lot of hard work, chicks finally break out of their shells. They stay close to the nest while they dry off.

Hens keep their brood together as they hatch. Each hen fluffs her chicks up and finds them a warm, safe place to rest a bit before they go discover the world.

Smart chicks

From day one, chicks can stand up and walk. They run around their mother, watching her peck and drink, and learning to do the same.

Chicks can feed themselves and they act like grownup chickens. The slightest noise scares them, however, sending them back to hide under their mother's wing.

Chicks are cute but also very fragile. Handle them with care.

Not all chicks are yellow. They can also be black, brown, or speckled.

Hens can find their chicks by the constant noise they make.

Crafty foxes find ways to gobble up chickens.

Sleeping chickens are not aware of predators in the coop.

Sometimes weasels get into a chicken coop and steal eggs.

Hens will run off with their chicks to escape danger.

Danger!

One of the many predators of chickens is the fox. Foxes eat chickens, usually by sneaking into a coop and snatching them while they sleep. Rats and weasels are known for stealing chicken eggs and chicks. Cats and dogs have also been known to catch chicks. It's important to lock up coops to keep chickens safe.

Growing up

Chicks grow up fast. After two weeks they have lost their yellow down and are getting feathers on their wings. They now have the coloring of a grown bird.

When they are a month old, you can tell if chickens are hens or cockerels. At this age young birds are ready to leave their mother and live as adults.

Once a chicken is grown, it's no longer afraid of other animals on the farm.

People and chickens

There are still some wild chickens living in the forests of India. In the United States, chickens are domesticated, meaning people take care of them. We raise chickens for food, for their eggs, and for their feathers.

Most chickens are raised on huge poultry farms.

The best chickens

There are many species of chickens. Some lay lots of eggs. Others are raised for food. There is even a species of rooster that is raised for its neck feathers, which are used in making fishing lures.

Hunting for eggs

Many people hunt for Easter eggs on Easter Sunday. Children search for Easter eggs, and the child who finds the most eggs wins a prize. Decorating Easter eggs is also a popular activity.

Chicks do not come out of chocolate eggs!

Battery chickens have access to food and water, but there is no hen to watch over them.

Faster and faster

Not all chickens are raised on farms. Some chickens are raised in a battery, a place where hundreds of them live together in a completely closed space. This method is used to produce more chickens faster. Battery chickens are force-fed pellets of food for 8 to 10 weeks before they are slaughtered and sold.

Other kinds of poultry

Other domesticated birds besides chickens live on farms. Each kind of bird lives separately and is not involved with the others. Most of these birds are from different families, but they are all poultry, meaning they are raised for eggs or meat.

turkey

Turkeys are native to North America. Males, called Toms, are impressive when they strut around and gobble in an effort to court females. Female turkeys, or hens, spend most of their time eating and waiting to be eaten!

ducks

Always quacking away, **ducks** are waterfowl, meaning they swim. Baby ducks, called ducklings, know how to swim at birth. Farmers raise ducks for their eggs, meat, and feathers.

geese

Geese are waterfowl, but they spend a lot of time in meadows eating grass. Geese are also raised by farmers for their feathers, eggs, and meat.

Guinea fowl are African birds noted for their alarm call. It lets farmers know there are people or animals trespassing on the farm. Guinea fowl are noted for their lean, flavorful meat.

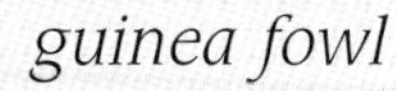

guinea fowl

You'll find the answers to these questions about chickens in your book.

Photograph credits:
COLIBRI Agency: R. Leguen: front cover, p. 15; D. Magnin: p. 2, p. 3 (bottom), p. 6, pp. 8-9, p. 11 (top right and bottom), p. 19 (top), pp. 20-21; D. Alet: p. 3 (top), p. 7 (top), p. 12; J.-Y. Lavergne: title page, p. 7 (bottom); P. Fontaine: pp. 4-5; É. Médard: p. 4 (bottom), p. 10; J. Negro-P. Cretu: p. 5 (bottom right); J. DuBois: p. 11 (top left); S. and J.-L. Ziegler: p. 13 (top); S. Bréal: p. 13 (bottom); C. Baranger: p. 14 (top left), p. 24 bottom; A. Labat: p. 16 (bottom); A.-M. Loubsens: p. 18 (top), p. 22 (bottom); J.-L. Paumard: p. 19 (bottom); G. Abadie: back cover, pp. 22-23; C. Testu: p. 23 (bottom); L. Chaix: p. 24 (top), B. Tauran: p. 25 (top); D. Fontaine: p. 25 (bottom)

JACANA Agency: H. Schwing: p. 5 (bottom left); Rouxaime: p. 14 (top right, bottom left, bottom right), p. 17; M. Claye: p. 16 (top); D. Cauchois: p. 18 (bottom left)